History of the 159th Regiment, New York State Volunteers

Compiled from the Diary of
Lieut. Edward Duffy

Originally Published in the *Hudson Gazette*, 1865.
One Hundred and Fifty Copies Reprinted New York, 1890.

Reformatted with a Table of Contents and Published by
Jessica Duke, Las Vegas, Nevada, 2011.

Book Layout and Cover Design by Jessica Duke.

Printed in the United States of America.
ISBN: 978-1-61393-050-2

Minor typographical errors have been corrected.

History of the 159th Regiment, New York State Volunteers

Contents.

History of the 159th Regiment, New York State Volunteers

During the latter part of October, 1862, negotiations were made by which the 167th Regiment, Colonel Homer A. Nelson, in Camp at Hudson, was consolidated with the 159th Regiment, Lieutenant-Colonel Edward L. Molineux, in camp at Brooklyn. The consolidated Regiment was designated the 159th, Colonel Nelson retaining command. The Regiment left "Camp-Kelly," Hudson, on the 30th day of October, proceeded on board the steamer Connecticut, arrived in New York next morning, and marched to Park Barracks. Remained there until November 1st, when we were mustered, into the United States service by Lieutenant R.B. Smith, U.S.A. Left Park Barracks and marched to Castle Garden; from there proceeded by steamboat to Staten Island, and went into Camp at New Dorp. Next day pitched our tents and had things very comfortable.

Colonel Nelson having been elected to Congress from his District, Lieutenant Colonel E.L. Molineux was appointed Colonel, and took command of the Regiment, which he virtually had from the first.

November 24th the Regiment broke Camp and was placed on board U.S. steam transport Northern Light, pier No. 3, North River, and remained at the wharf until December 2d, when we hauled into the stream. Early on the morning of the 4th weighed anchor, and the 159th Regiment put to sea. On the 13th we reached Ship Island, in the Gulf of Mexico, having enjoyed a tolerable good passage for the season of the year, being more fortunate than other ships of the expedition, some of them having suffered considerable from rough weather off Cape Hatteras.

December 14th reached New Orleans, and anchored in the stream over night.—The following day pursued our course up the river to Baton Rouge, and arrived there on the 17th. The enemy, learning of our approach in force, concluded to evacuate, while our monitors gave them a parting salute, and the same day the Stars and Stripes were hoisted to the breeze from the Capitol, amid the shouts and cheers of the gratified soldiers.

Now the work began of making thorough soldiers of men, the greater portion of whom never used fire arms before, at least not in the manner required by the service. Squad, Company, Battalion, and Brigade drill, with any quantity of discipline considered essential to fit men for the campaigning and hardships visible in the distance, were gone through with.

Perhaps few in the volunteer service, none of whom could boast of very much practical experience, were better adapted than Colonel Molineux for this severe task; very quick, energetic, ambitious to do his own duty and to keep every man in his command busy, was the true secret of his success as a disciplinarian.

For nearly three months the men were kept steadily under instruction, and became quite proficient in the use of the musket, and all the essential discipline to make an effective army.

On the 13th of March, 1863, broke camp, and the army moved up to the rear of Port Hudson. Colonel Molineux having command of a provisional Brigade with Nims' Massachusetts Battery, went up the Clinton Road, while the main army proceeded down the Port Hudson Road about eighteen miles, skirmishing the Rebels the whole way, driving their pickets and scouts as they advanced.

At this time Port Hudson was strongly manned, there being from 23,000 to 25,000 men in that natural stronghold. Maneuvered about this quarter until the 20th, when we again joined the main body of the army on the Port Hudson Road, returning to Baton Rouge, Louisiana.

This movement was made to attract the attention of the enemy, and enable Admiral Farragut's boats to proceed up the river past the fort that here impeded his passage. This was a bold but brilliantly successful move, that only an "iron Farragut" could have accomplished. This blind enabled General Banks to more successfully pursue his future designs, as the enemy had been led to suppose by the formidable movements around Port Hudson that a general attack was to be made at once to reduce the place. Subsequent events exhibited the picture in a different light.

From the 20th to the 28th of March we were kept in readiness to move at a moment's notice. Finally, the suspense was removed and we proceeded on board the transport ship Laurel Hill, to Donaldsonville, La., where we landed in a drizzling rain, about 10 o'clock, P.M., with mother earth for a couch and the broad, moist sky for a canopy. Active campaigning was now fairly inaugurated.

On the 31st of March the troops moved for Thibodeaux, La. The 159th was detailed in charge of supplies and Regimental property, and proceeded by boat up the Bayou Lafourche, arriving at Thibodeaux April 1st. On the 3d we moved to the Railroad Station at Terra Bone, taking the cars for Bayou Bueff, where we arrived on the 4th. Remained here until the 9th. Arrived at Brasher City, La., on the 11th, in company with the 13th Connecticut, 26th Maine, and a detachment of Cavalry. Boarded river steamer Laurel Hill, and proceeded up Berwick Bay, into Grand Lake, accompanied by Grover's Division, numbering about 8,000 men. Had with us three small gun-boats, moving cautiously. Reached the Bend without disaster, the gun-boats shelling the woods and covering the landing of the troops, which took place on the morning of the 13th. Skirmishing lines were thrown out immediately. The Lake is three or four miles wide at this place, and is called Shell Bend. Having all ashore and everything in good order, the enemy in sight, we retired for the rest we needed, little realizing the hard fight before us. Lay in skirmish line all night. A few of our men were wounded.

At early dawn of the 14th of April, the lines advanced without breakfast, marching about a mile and a half. The enemy was strongly posted in a wood at a bend in the Bayou. Covered by the gun-boats, the 25th and 13th Connecticut and 26th Maine, commanded by Colonel Birge, were skirmishing briskly in front. Colonel Molineux was ordered to take his Regiment, the 159th, and advance and charge the woods.

We advanced in good style over a plowed cane-field in line, passing over the 26th Maine, who were lying down. Passed the skirmish line of the 25th Connecticut, who were under cover of the cane on our right. Several of our men fell in the advance. Reached within pistol shot of the fence and wood where the enemy was concealed. Scarcely one could be seen while they poured a most effective fire on us, but we steadily advanced 'till ordered to lay down. At this time the men were nearly exhausted, marching at double quick over rough ground with heavy knapsacks; it took a little time to catch fresh wind and unburthen ourselves of our heavy load. We could not have stopped at a more uncomfortable place, for the enemy gave it to us hot and sweet, while we did not have a chance to see them. They came out of the wood through the cane to the rear of our right flank, and right on top of us. We no doubt would have layed there 'till every man of us was shot had not the order come to fall back to the left. Several of our men were taken prisoners, the enemy rushing upon us while rising up from our position, and poured a most deadly fire into us with fearful effect. The 91st N.Y.S. Volunteers coming down to our aid, the rebels skedaddled, but not without some loss and a number taken prisoners.

Col. Molineux was severely wounded in the mouth, Lieut-Col. Draper and Adjutant Lathrop were killed; the Colonel, Lieut-Colonel and Adjutant were nobly doing their duty in the advance, leading their men. No officers could have done better or been more brave. They were picked out by the enemy's sharpshooters posted in the trees near by.

The victory was ours, though the Regiment paid dearly for this, their maiden fight. Second Lieutenant Lockwood, of Company G, was killed while nobly leading his Company. Lieutenants Plunket and Price were mortally wounded. Lieutenant Manley, of Company A, was killed, and Lieutenant Tieman and Captain Petit were slightly wounded. Our total killed, wounded and missing amounted to 112, viz: 6 officers, 23 men killed; 2 officers, 69 men wounded, and 12 men prisoners. Major Burt, who was on General Grover's staff, now assumed command of the Regiment.

The 91st N.Y.S. Volunteers were to have advanced with us on the right, but misunderstanding the order, they failed to advance, causing us to be flanked and receive a heavy enfilading fire by which we suffered so much.

Our Division was to fall on the rear of the enemy, when they were driven from their entrenched position, at Bislin, south of Franklin. The strong resistance at Irish Bend was to make good their escape, which they effected at the loss of a large number of prisoners.

April 15th, marched up Teche 18½ miles, the main column, under General Banks, in advance. Arrived at New Iberia, where Mills and Ashton, of Company K, who were taken prisoners at Irish Bend, joined the Regiment, the enemy having paroled them.

Arrived at Vermillion Bayou, and finding the bridges burned, had to reconstruct them. The Regiment was now detailed to collect cattle through the prairie and drive them to Berwick City. We collected about three thousand head.

A detachment of the Regiment left us to gather up cotton and other property laying about loose. Arrived safely at Berwick City, and returned in charge of a wagon train which we left at Opolosus, and reported to the Division Commander at Barrie's Landing, on the Teche, eight miles from Opolosus.

May 5th broke camp and marched to Little Washington, La., and from there to Wells' Plantation, where we went into camp. Left Welles' Farm and marched to Simsport, a distance of eighty miles, where we arrived on the 18th, and crossed the Mississippi, landing at Boyou Sara, on the night of the 21st.

The 24th day of May brought us close to the enemy's outer works in front of Port Hudson, after marching the distance of eight hundred miles from the 28th day of March to the 24th day of May.

Our position now placed us under a heavy fire and shelling from the fort. The enemy well knowing the Road we were obliged to advance on, poured an accurate fire upon our line. But few casualties occurred, although some narrow and hair-breadth escapes happened.

On the 25th of May, skirmishing all day. Result, four of our men killed. At noon our men were relieved from picket, and the Regiment ordered to the right of our Division. A general movement was made along the lines, and our Regiment was selected to attack a portion of the enemy's works, and storm it. The 25th Connecticut Volunteers was consolidated with us, commanded by Major Burt. It was necessary to make a circuitous route three miles through the woods to the right, to reach the position to be attacked, exposed the whole way to a continued and terrific fire of shot and shell; but our boys unflinchingly pressed on through ravines, over felled trees, and all sorts of intricacies natural and artificial. The final assault was to be made upon an almost perpendicular slope. "Forward!" was the word, and persistently we advanced, reaching just under and near the parapet, but the fire was like hail; the Color Bearer was shot dead and the color staff shot from his hands, but it was again secured and brought off. We lay in this position for some hours unable to advance or retreat; it seemed almost impossible for one to escape under such a fire. A number of our men remained in this position until after dark, when the firing ceased. Shortly after midnight, the enemy supposing we still lay close to their works, sallied out and poured a heavy volley into the position from which we had been very prudently removed but a short time previous. We captured one Captain and eight sharpshooters in ambush outside the works; this was but little, yet it furnished some satisfaction for our loss. This was in advance of any previous attack, several of which were made during the day. Our loss on this eventful day was 21 men killed and 38 wounded. From this time until June 14th we were almost continually in the rifle pits.

June 14th was selected for a general assault. The advance stormers, led by the gallant Colonel Paine, of the 4th Wisconsin Volunteers, who had been acting Brigadier General for some time previous, pressed on under the most severe fire. A number succeeded in penetrating the enemy's works, but owing to the obstructions we were obliged to pass over, the advance could not be supported with the necessary rapidity for the success of the scheme. Colonel Paine being severely wounded early in the action, materially injured the success of the enterprise. The wounded Colonel lay in such a position that he could not be removed until after dark; several attempts were made but the parties were either killed or wounded in their noble efforts. It was in this engagement that the gallant Colonel Cowles, of the 128th, lost his life while leading his men to the assault.

Shortly after midnight we left the pits where we had been for several days, to join the column of attack coming up at daylight, having to defile through the woods several miles. General Grover's Division supported the advance. The 159th advanced under a severe fire through a ravine and over obstructed ground to a commanding position, a knoll overlooking the enemy's works; here we lay in position until between three and four o'clock, P.M., the enemy firing a continuous volley over our heads. No thanks to them that our craniums escaped. It was contemplated to make a second assault, and we were ordered to the left, some distance over clear and exposed ground to join the forces in waiting for this purpose. We remained here until after dark, and the firing having ceased, further attempts were deferred, and we moved back to the position we gained in the morning, and were set to work fortifying, but were soon relieved by colored entrenchers, and returned to our quarters in the woods, which we reached in the small hours of the morning, greatly fatigued. Our loss on this occasion was not severe; 12 men wounded.

To reduce the place was the work assigned, and it must be accomplished. General Banks issued an order on the 15th of June, congratulating the troops for their behavior and close investment of the stronghold, and calling for 1,000 Volunteers from the forces to form a storming column or "a forlorn hope." Soon more than the required number were on hand, and formed into two Battalions, to be commanded by General Birge. It did not become necessary to make this assault. General Gardner hearing of the fall of Vicksburg, capitulated on the 8th of July.

Every thing being satisfactorily arranged, our forces, preceded by the storming party, entered the fortification and filed past 6,000 brave but discomfited "Gray Backs;" freedom's emblem, the Stars and Stripes, was soon hoisted, saluted by a discharge from the guns which had so recently belched forth death against our lines.

On the 11th of July we proceeded down the river to Donaldsonville on board the steamer Iberville. The enemy a few nights prior to the surrender, made a desperate attack on a small garrison in the fort at this place, but were repulsed with severe loss. The garrison numbered not more than four hundred; more than three hundred of the enemy were seriously wounded. The enemy was posted just behind the town; batteries were placed along the levee at numerous places; several boats had been destroyed, and the transportation of supplies was getting quite precarious, but the surrender of Port Hudson put a stop to their amusement. We landed at night, slept on our arms, and woke up in the morning close to the enemy's pickets.

On the 14th a Brigade commanded by Colonel Morgan, of the 90th N.Y. Volunteers, advanced upon the Bayou about four miles, driving the enemy before him. The 159th was on his right flank doing picket duty, and the Company I belonged to was on the outside post in command of Captain William H. Sliter. Colonel Morgan came up to us and ordered us to go with him. The Captain told him he would not leave his post, a most important one, that the whole Brigade depended upon.

On the fifteenth the enemy made a stand under cover of a thick wood, protected by heavy artillery. Finding our forces not very formidable, the enemy advanced in force on our left flank, taking a number of prisoners. Reinforcements at this time came up, and the enemy fell back west of the Atchafalya River.

July 16th the 159th was detailed to guard wagon trains on the west side of the Mississippi. Arrived at Carrolton, where we were allowed to rest, remaining until the last of August, when we were sent to Thibodeaux, La., via Algiers.

September 1st, reached Thibodeaux. General Birge was in command of the District of Lafourche. Our Regiment, with the 13th Connecticut, was detailed to do provost and picket duty, while the other troops were distributed over the District.

Colonel Molineux was appointed on General Franklin's Staff on the 24th of September. We remained at this position until March 18th, 1864. Quiet prevailed during our advent here, only a few night alarms occuring, causing the long roll to beat and the men to turn out, but they amounted to nothing serious.

January 1st, 1864, Lieutenant-Colonel Burt took command of the Regiment at Thibodeaux, Colonel Molineux was relieved from duty on General Franklin's Staff, and assigned to command the Lafourche District, in place of General Birge, relieved on furlough.

On the 7th, Lieutenant-Colonel Burt resigned his Commission, on Surgeon's certificate, and was honorably discharged, and the command devolved on the senior officer, Captain Hart. His reign, however, was short. Major Gaul, who was on detached service at Albany, N.Y., was appointed Lieutenant-Colonel, vice Burt, and Captain Waltermire made Major. This arrangement was highly satisfactory to the whole Regiment.

February 25th, Lieutenant-Colonel Gaul reported for duty and took command of the Regiment.

A new Company of 64 enlisted men arrived from Hudson, N.Y., under command of Captain James S. Reynolds, First Lieutenant E. Spencer Elmer, and Second Lieutenant Peter R. Van Deusen. Old Company G was broken up, the men distributed through the Regiment, and the new Company installed in their place.

On the 19th of March we broke camp for the spring campaign, having been here nearly seven months. On the 21st took cars for New Orleans. Arrived at Algiers on the 24th. Embarked on board the James Battel and arrived at Alexandria, La., via. Red River, on the 27th of March.

Marched to west side of the river to Pineville while the dam was being built to enable the gunboats to pass down the river. Colonel Molineux was relieved from command of a recruiting party which he had been in charge of, called the "Louisiana Scouts," but the Regiment nick-named them the "Jay-hawkers." The gunboats having safely passed the dam, the army commenced moving back on the Mississippi.

May 11th, the 159th with some artillery and cavalry, were placed in charge of Alexandria and defences, under Colonel Molineux, and remained there while General Banks moved on Shreveport.

The engagements of Grandecore, Sabine Cross Road and Cane river, occurred while we were here. General Banks not being able to keep up his supplies, as the gunboats could not pass up in consequence of the rapid falling of the river, was obliged to fall back.

On the 16th, while passing through Marksville, the enemy made considerable show of resistance. The union forces deployed in line, making a grand and imposing appearance, extending for several miles over an almost level plain. The artillery on both sides belched forth for some hours. The casualties were light, and the enemy driven back. This is called the "Battle of Mansura."

On the 19th, reached Simsport; this was our second advent here.

The enemy continually harrassed us from the time we left Alexandria, from across the rivers and Bayous, and on our flanks and rear, but accomplished but little damage.

Crossed the river on the steamer Cumberland, and reached Morganzi, La., on the 22d. This terminated the "Red River Expedition" of 1864.

June 19th, General Grover's Division proceeded up the river as far as Fort Adams, and scoured the banks on either side for guerillas, who were numerous, firing into all boats passing on the river. Captured a few prisoners and returned. July 2d, Lieutenant-Colonel

Gaul resigned, on account of disease contracted in the campaign, and Major Waltermire was promoted to Lieutenant-Colonel.

July 3d, took steamer Lancaster, and arrived at Algiers, opposite New Orleans, at 6 P.M., July 4th.

July 17th, went on board the U.S. Transport Cahawba. At 12 o'clock that night moved down the river and arrived at the Rip Raps and Fortress Monroe, on the 24th. Received water, and on the 25th proceeded up the James river, arriving at Bermuda Hundreds at 5 P.M. Move up to the entrenched position, and were kept continually moving about while there.

August 1st, went on board the steamer Winona, and arrive at Washington, D.C., the next day. Put baggage on cars for Harpers Ferry, but orders countermanded before we got off. Marched through Washington to Tenallytown. Remained there until the 14th, when we started to join General Sheridan in the Shenandoah Valley, through Snicker's Gap. Crossed Chain Bridge and encamped at Owl Run, Va., that night. Arrived at Leesburgh on the 17th; passed through Hamilton, and within four miles of Snicker's Gap. Here a dispatch notified us that the enemy was hurrying to cut us off at the gap. This notice was timely, and saved us a serious disaster. Immediately moved on, forded the Shenandoah river, marched nearly all night, and reached Sheridan's forces on the morning of the 18th, having marched about forty miles the previous day and night.

Fighting had been going on in that vicinity for some time before we arrived. Were ordered to throw up temporary breast-works, which was quickly accomplished.

Early in the morning of the 21st, heavy artillery firing was heard on our right. The 6th and 8th Corps were engaged in a heavy battle that day, and late in the afternoon our Division was moved to the right of the 6th Corps and in front of Charlestown. In this engagement the loss was heavy on both sides.

General Sheridan then drew his forces back to the rear of the defences of Harper's Ferry, at Halltown, the 19th Corps covering the movement. Our Regiment was put on picket duty behind Bolliver Heights, and a constant picket firing was kept up on both sides.

On the 24th, our Regiment, with the 22d Iowa and 11th Indiana, under command of Colonel McCauly, advanced on the skirmish line to reconnoiter the enemy. Drove them back some distance, advancing in good style under a heavy fire, and maintained our position until ordered to retire. We were under a heavy artillery fire for about two hours, and our Regiment lost one officer and twelve men.

Two officers and sixty picked men were selected to join others to advance for the purpose of bringing on an engagement, but news came that the enemy had retired. The cavalry followed them, and occupied Charlestown.

A grand advance was ordered, and on the 3d of September we started off with four days' rations in our haversacks.

Advanced to near Berryville, where heavy artillery firing was heard in front, soon followed by musketry, gradually growing more rapid. The enemy had gained a slight advantage on the left of the 8th

Corps. The 19th Corps moved quickly to their support, when the enemy fell back, but firing continued until after dark.

On the morning of the 5th, three lines of rude breastworks were thrown up in double quick time—hand, feet, bayonets, tin cups, old shoes, every thing was brought into requisition to accomplish the work, which was completed during a heavy rain.

On the 6th, the enemy fell back to the Opequan Creek.

On the 7th our Regiment proceeded to the Opequan, reconnoitered the enemy, and returned the same day after accomplishing our objects.

On the 11th captured the 8th South Carolina Regiment, with all its officers.

September 17th, General Grant made a short visit to "Little Phil," which was set down as indicating hard work ahead, in which supposition we were not disappointed.

At 1 o'clock, A.M., on the 19th, a general movement of the whole army began. Skirmishing towards Opequan Creek became more and more brisk, till it assumed all the proportions of a fierce battle, lasting the whole of the day. Alternately the opposing forces were repulsed in turn, either side contesting for the superiority with the most dogged persistency. Only the ability and determination of the gallant "Little Phil." could have secured success. We had 5 men killed, 4 officers and 36 men wounded, and 1 officer and 20 men taken prisoners.

The enemy was closely followed up to Fisher's Hill, behind Strausburg, a well fortified and naturally strong position.—Gaining the point we desired on the 22d, it was determined to force the enemy up the valley, and occupy this stronghold.

The 6th Corps was thrown around to the left and rear of the enemy by the base of the mountain; this movement took almost all day. The 8th Corps was on the left, and the 19th in the centre. When the 6th Corps reached sufficiently near, a grand movement was made, our boys forcing the centre. This action took the enemy by surprise and they retreated up the valley in tall style. About a thousand were not able to carry out their intentions, and it devolved on us to pilot them to the rear.

The chase was kept up all night, and we reached Woodstock the next morning. We suffered no loss on this occasion. It was indeed a cheap victory. We captured a number of horses and wagons, artillery, and any quantity of small arms, which our Regiment was detailed to take charge of and convey to Winchester, with the prisoners.

We left Woodstock at five P.M., and delivered the property and prisoners at Winchester, and on the 25th started for the front again, in charge of a supply train.

On the 27th, at three P.M., arrived at the front at Harrisonburg, having marched over one hundred miles in less than four days.

On the 30th, the 6th and 19th Corps advanced to Mount Crawford, the enemy showing some disposition to interrupt the Cavalry. Nothing serious being discovered, we fell back to Harrisonburg.

Remained here until October 6th, when we moved back to New Market, and on the 9th arrived at Woodstock. Marched 'till 9 A.M., arriving a short distance south of Fisher's Hill. The enemy finding us falling back, closely followed after. We were thrown into line on either side of the road, ready for what might occur. A little skirmishing with the rear guard was the only demonstration, and at four P.M. we were back in front of Fisher's Hill, our old position. On the 9th, General Rosier, with his artillery and cavalry, hovered about our rear, being closely watched by our cavalry. He came a little too near, however, and our cavalry dashed at him and captured seven or eight guns and a number of prisoners.

On the 11th of October, marched back to the north side of Cedar Creek, which we commenced fortifying. The enemy brought heavy Batteries and shelled the 8th Corps camp on the left. The trains were sent to the rear, and the troops placed in line ready for action, but the enemy appeared to be reconnoitering, and fell back to Fisher's Hill.

All remained quiet until the morning of the 19th of October. Early had received large reinforcements from Richmond, and now made a last desperate effort to redeem his lost laurels in the valley. It was a well executed and daring move, and for a time promised success. He moved his men during the night around our left flank by the base of the Blue Ridge, in single file, many not even carrying their canteens, fearful that the least noise would be made. In this manner they succeeded in reaching Middletown, a mile and a half in the rear of our breastworks; before daylight a feint was made on our right to attract our attention in that quarter; a short time after a volley or two of musketry was heard on our left, the enemy dashing on the 8th Corps in desperate fury, completely surprising them. So sudden was the attack that many were captured before they had time to leave their tents or seize their muskets. On pressed the successful mass, shouting and yelling in the wildest manner.

The 8th Corps, badly demoralized, poured back on the rear of the 19th Corps, closely pursued by the enemy. Our Division was going out to reconnoiter, and were in line, but from their position could do but little, the enemy being in our rear, so that not a shot could be fired without danger to our own men. The 1st Division, 19th Corps, was sent to support the 8th Corps early in the morning, and suffered severely, meeting the first onslaught of the enemy. Our Division (the 2d) took position in front of the breastworks, but being of no service there, we filed to the right and fell back to the rear where we could be re-formed and occupy a position in front of the enemy.

The enemy steadily pressed us back four or five miles. Matters began to look blue, when the dashing "Little Phil" came up as fast as his noble black steed could carry him, leaving his attendants far in the rear. The noise of the battle had reached him at Winchester early in the morning. The appearance of Sheridan immediately instilled new vigor, energy and determination into the men. He passed along the whole line amid the most marked enthusiasm, telling the men they would quarter in their old camp again that night.

The broken lines were speedily re-formed, the General passing along hat in hand, encouraging the men. This was sufficient, and from this dates the last advent of Early in the valley.

It was now our turn. The enemy charged us, and for the first time were repulsed. We pressed on determined to win. The success of the morning turned to a most irretrievable and disastrous defeat to the enemy. They were completely routed, suffering a terrible slaughter. Twenty-four guns captured in the morning were retaken, besides a large number of prisoners, and most of the enemy's artillery, numbering over fifty pieces. Our Regiment took 16 officers and 34 men as prisoners, in this engagement. We lost Captain Richmond, one of the best officers in the Regiment, and a brave, noble fellow. He was shot in the afternoon, when success began to turn on our side. None braver paid the penalty of death for his country. We had 2 privates killed, 10 wounded, and 5 taken prisoners.

The cavalry pressed the beaten foe until horse flesh could do no more, taking a large number of prisoners and all sorts of war implements and materials. Thus was this long day spent in fighting and running, advancing and retreating, now one side victorious, then the other, when finally success crowned our efforts.

Major Hart, of our Regiment, on General Grover's Staff, was wounded and taken by the Rebels. He was not attended to in time, and lost so much blood as to cause his death.

On the 20th, the forces moved about three miles up the valley, overlooking Strausburg, the cavalry continuing the pursuit to Harrisonburg, capturing more artillery and wagons. On the 21st moved back to our old position on Cedar Creek.

From this time until the 1st of January, 1865, the men were engaged in erecting breastworks, preparing Winter quarters, frequently moving and occasionally skirmishing with the enemy along the lines.

The 24th of November was observed in camp as Thanksgiving Day, and all duties were stopped that could be dispensed with. Thanks to our kind friends at home, we were provided with a bountiful feast of turkeys, chickens, pies and other luxuries, and if they could have witnessed the satisfaction of the men on that occasion, it would have been ample reward for their generosity.

January 6th, 1865, moved to Harper's Ferry. Arrived in Baltimore next morning and quartered in Barracks on Carrol Hill. On the 11th Colonel Waltermire took command of the Regiment, and we embarked on board the steamer Sua-Noda, for Savannah. General Grover and Staff, the 128th N.Y.S. Volunteers and the 24th Iowa were on the same vessel.

On the 18th cast anchor in Warsaw Sound, eight miles from Savannah; and on the 20th the Regiment went up to the city on river boats, and were quartered in the Central Railroad Depot.

On the 26th were moved out to the fortifications, on the West side of the town.

February 1st, fresh bread was issued with our rations, which was a luxury to the boys so long kept on "hard tack." February 19th, fired a rousing salute on hearing of the occupation of Charleston by the Union forces. On the 22d, celebrated Washington's Birth-day in a becoming manner.

March 9th, were ordered on board the Tug boat U.S. Grant, which conveyed us to Hilton Head, where we went into Barracks.

On the 15th, were taken on board U.S. Transport New York, a splendid new ship, and arrived at Charleston, S.C., at one A.M., on the 16th. On the 17th took on board the 52d Pennsylvania, a detachment of the 54th New York, and the 28th Iowa, in all about 1,600 men. Weighed anchor on the 18th at ten and a half o'clock A.M., and moved down the harbor. This gave us an excellent opportunity to see the dilapidated city and its approaches, fortifications and defences; the latter of which were exceedingly formidable, and might be considered impregnable from the water side.

March 19th, anchored off Fort Fisher, at nine o'clock A.M., when we received orders to report at Morehead City, N.C. Reached that port on the 20th, landed on the 21st, and awaited orders.

April 5th, intelligence reached us of the evacuation of Richmond and Petersburg, which caused great rejoicing throughout the camp. This was followed by the more encouraging news of Lee's surrender on the 9th. While these great victories were being celebrated, the sad intelligence of the assassination of President Lincoln reached camp, and cast a deep sadness over those who had been jubilant but the hour before.

May 3d, ordered to report back to General Grover, at Savannah. Break camp, and embarked on board steamer Star of the South. On the 7th, after an eventful trip, disembarked at Savannah, and found the City remarkably improved in appearance since we left it.

May 11th, ordered to proceed to Augusta. Took up our line of march in a heavy rain storm, and made twelve miles that day through the woods. Next day we accomplished over twenty miles.

On the 14th an Orderly from General Molineux' Headquarters reached us, to hurry up our march. The 159th, 128th and 131st N.Y.S. Volunteers in advance of all, to make Wainsborough and take the cars. Reached Augusta on the 17th, pretty well used up from fatigue.

June 7th, a general review of all the troops by General Molineux, on which occasion he issued a congratulatory order to the soldiers, complimenting them for their excellent discipline, and the services they had rendered.

Here the Regiment virtually closed its campaign, nothing further of note occurring up to the present writing, beyond the usual routine of camp life in the city.

List of Field, Staff and Line Officers of the 159th Regiment at date of muster into the United States Service, November 1st, 1862:

Homer A. Nelson	Colonel, discharged.
Edward L. Molineux	Lieutenant-Colonel, promoted to Colonel, breveted Brigadier General, discharged.
Gilbert Draper	Major, promoted to Lieutenant-Colonel, killed at Irish Bend.
Robert Lathrop	Adjutant, killed at Irish Bend.
Charles A. Robertson	Surgeon, discharged.
William Y. Provost	First Assistant Surgeon, promoted to Surgeon, discharged.
Caleb C. Briggs	Second Assistant Surgeon, promoted to Surgeon.
Mark D. Wilber	Quarter-Master, discharged.

Company A

Edward L. Gaul	Capt., promoted to Major and Lieutenant-Colonel, discharged.

Edward Atwood	First Lieut., discharged.
Wesley Bradley	Second Lieut., died of fever.

Company B

A.J. Dayton	Capt., discharged.
Harry Tiemann	First Lieut., discharged.
Alfred Greenleaf	Second Lieut., discharged.

Company C

Ariel M. Gamwell	Capt., discharged.
Crawford Williams	First Lieut., discharged.
Edward Hubbel	Second Lieut., discharged.

Company D

Jacob Hattry	Capt. Co. D, discharged.
Lawrence Lorette	First Lieut., discharged.
John Manly	Second Lieut., promoted to First Lieut., killed at Irish Bend.

Company E

William Waltermire	Capt, promoted to Major, Lieutenant-Colonel and Colonel.
Nathan S. Post	First Lieut., discharged.
Robert Traver	Second Lieut., discharged.

Company F

Robert McD. Hart	Capt., promoted to Major, killed.
William Burtis	First Lieut., discharged
George W. Hussey	Second Lieut., promoted to Capt.

Company G

William Sliter	Capt., discharged.
Charles Lewis	First Lieut., promoted to Colonel of the 176th Regiment.
Byron Lockwood	Second Lieut., killed at Irish Bend.

Company H

Welles O. Petit	Capt., promoted to Major.

Charles C. Baker	First Lieut., promoted to Captain.
George R. Herbert	Second Lieut., detached.

Company I

Edward Wardle	Capt., discharged.
John W. Shields	First Lieut., discharged.
Jacob Fingar	Second Lieut., discharged.

Company K

Joe B. Ramsden	Capt., discharged.
William Plunket	First Lieut., killed at Irish Bend.
Duncan Richmond	Second Lieut., promoted to Captain, killed.

Promoted from the Ranks.

William F. Tiemann	to	Captain
John H. Charlot		Quarter-Master
Edward Tynan		First Lieutenant
Barzilla Ransom		First Lieutenant
Henry M. Howard		First Lieutenant
Christopher Branch		First Lieutenant
Alfred Bruce		First Lieutenant
Lambert Dingman		First Lieutenant
Andrew Rifenburgh		First Lieutenant
Edward Duffy		First Lieutenant
E. Parmley Brown		First Lieutenant
John Day		First Lieutenant
John A. Tiemann		First Lieutenant
M.A. Dunham		First Lieutenant

William Spanburgh	Second Lieutenant
Charles P. Price	Second Lieutenant
Herman Smith	Second Lieutenant

List of Battles and Skirmishes in which the Regiment was engaged.

Irish Bend, La., April 14th 1863, killed, 6 officers, 23 enlisted men; wounded 4 officers, 69 enlisted men; prisoners, 12. Total loss—112.

Before Port Hudson, La., May 26th, 1863, killed, 4 enlisted men.

Port Hudson, La., first assault, May 27th, 1863, killed, 21 enlisted men; wounded, 38. Total loss—59.

Port Hudson, second assault, June 14th, 1863, wounded, 12 enlisted men.

Mansura, La., May 16th, 1864. No casualties.

Halltown, Va., August 24th, 1864, killed, 1 enlisted man; wounded, 1 officer, 10 enlisted men; prisoners, 1 enlisted man. Total loss—13.

Berryville, September 3d, 1864, killed, 1 enlisted man; wounded, 2 enlisted men. Total loss—3.

Opequan, Va., September 19th, 1864, killed, 5 enlisted men; wounded, 4 officers, 56 enlisted men; prisoners, 1 officer, 20 enlisted men. Total loss—86.

Fisher's Hill, Va., September 22d, 1864. No casualties.

Cedar Creek, Va., October 19th, 1864, killed, 2 officers, 2 enlisted men; wounded, 1 officer, 10 enlisted men; prisoners, 5. Total loss—20.

General Officers under whom the Regiment served during the war.

Generals

- Banks,
- Grover,
- Auger,
- Reynolds,
- Emory,
- Birge,
- Sherman,
- Schofield,
- Terry,
- Gilmore,
- Thomas,
- Sheridan,
- Steedman,
- Wright,
- Canby,
- Birney,
- Molineux,
- King.

List of Cities and Towns the Regiment has visited in the line of military service.

Louisiana

Baton Rouge,	Donaldsonville,	Carrollton,
New Orleans,	Algiers,	Terra Bone,
Thiladuex,	Brashar City,	Bayou Bueff,
Berwick City,	Franklin,	New Iberie,
Vermillionville,	Washington,	Bears Landing,
Opolosus,	Chaneyville,	Simsport,
Bayou Sara,	Port Hudson,	Clinton,
Alexandria,	Pineville,	Patersonville,
Mansura,	Williamsport,	Morganza,
Point Coupee,	Teunice Bend,	Jefferson.

Virginia

Bermuda Hundred,	Leesburgh,	Snickerville,

Castle Burough, Berryville, Charlestown,

Halltown, Harper's Ferry, Winchester,

Kernstown, Newtown, Middletown,

Strasburgh, Edenborough, Newmarket,

Mount Jackson, Harrisonburgh, Mount Crawford,

Centerville, Stephenson Station, Burseville.

Georgia

Savannah, Alexandria, Waynesborough,

Allen, Green, Bashaw,

Augusta.

District of Columbia

Washington, Georgetown, Tenallaytown.

North Carolina

Wilmington, Morehead City.

South Carolina

Hilton Head, Charleston.

Mississippi

Fort Adams.

Maryland

Baltimore.

River Transportation during the War.

BOAT	FROM	TO
1863.		
St. Mary	Baton Rouge	Donaldsonville.
Empire Parish	Donaldsonville	Thibadaux.
Laurel Hill	Brasher City	Irish Bend.
Empire Parish	Symsport	Bayou Sara.
Laurel Hill	Port Hudson	Donaldsonville.
Gen. Banks	Carrollton	Algiers.
1864.		
James Bartlet	Algiers	Alexandria.
Ohio Belle	Morganza	Fort Adams.
Lancaster No. 3	Morganza	New Orleans.
Wynonah	Bermuda Hundred	Washington.
1865.		
Clifton	Warsaw River	Savannah.
U.S. Grant	Savannah	Hilton Head.
H.M. Wells U.S.T.	New York	Morehead City.

Sea Voyages of the Regiment during the War.

BOAT	FROM	TO
1862.		
Northern Light	New York	Baton Rouge.
1864.		
Cohola	New Orleans	Bermuda Hund.
1865.		
Suwo-Noda	Baltimore	Savannah.
New York	Hilton Head	Morehead City.
Star of the South	Morehead City	Savannah.

List of Field, Staff and Line Officers now in command of the Regiment.

Colonel	WILLIAM WALTERMIRE.
Major	WELLS O. PETIT.
Acting Adjutant	GEORGE B. STALEY.
Surgeon	CALEB C. BRIGGS.
Acting Quarter-Master	E. SPENCER ELMER.
Company A.	Capt. William F. Tiemann.
Company B.	First Lieut. John Day.
Company C.	First Lieut. Barzilla Ransom.
Company D.	First Lieut. E. Parmley Brown.
Company E.	First Lieut. Andrew Rifenburgh.
Company F.	Capt. George W. Hussey.
Company G.	Capt. James S. Reynolds.
Company H.	-----

Company I.	First Lieut. Edward Tynan.
Company K.	First Lieut. E. Spencer Elmer.